The World's Harvest

ENERGY from SUN, WIND, and TIDE

Jacqueline Dineen

The exciting story of the power being harnessed today and in the near future, from sources which will never run out.

ENSLOW PUBLISHERS, INC.
Bloy St & Ramsey Ave.
Box 777
Hillside, N.J. 07205

The picture above shows a vertical axis wind turbine in the United States.
[*cover*] *The cover picture shows the solar power station at Odeillo in the French Pyrenees.*
[*title page*] *The picture on the title page shows a wave breaking on a beach, wasting energy!*
[1–26] *All other pictures are identified by number in the text.*

Contents

This series was developed for a worldwide market.

First American Edition, 1988

Printed in the United States of America

10 9 8 7 6 5 4 3 2 1

LIBRARY OF CONGRESS
Library of Congress Cataloging-in-Publication Data

Dineen, Jacqueline.
Energy from sun, wind, and tide / Jacqueline Dineen.
p. cm. -- (The world's harvest)
Includes index.
Summary: Describes how the limitless supplies of power derived from the sun, wind, and tide are being harnessed and explains their importance in a world where traditional energy supplies are becoming scarcer.
ISBN 0-89490-214-8
1. Renewable energy sources--Juvenile literature. 2. Solar energy--Juvenile literature. 3. Wind power--Juvenile literature. 4. Tidal power--Juvenile literature. [1. Renewable energy sources.] I. Title. II. Series: Dineen, Jacqueline. World's harvest.
TJ808.2.D56 1988
621.4--dc19 88-3549
CIP
AC

Introduction

We have not always had electricity, but we have had it for long enough to grow used to it. How would you like to go back to the days of candles and oil lamps, and cooking over an open fire? It might be fun for a day or two, but after a while I think it would make life difficult. Electricity has become very necessary in the modern age.

Electricity is made by turning mechanical energy into electrical energy in a machine called a generator. The mechanical energy comes from burning coal and oil in power stations, like the one in picture [1].

The problem is that coal and oil won't last for ever. We are using them up so quickly that they will run out one day, and then what will we do? There will be no electricity unless new

A power station is a factory for generating electricity.

[1]

sources of energy can be found.

In this book, I tell you something about the sources of energy that might soon be replacing fuels. First, I explain what energy is and how electricity is produced at the moment. Then I talk about the 'renewable' energy sources.

Renewable energy sources are those which renew themselves, and can never run out. However much energy we take from the sun, the wind, and the tide, there will always be further energy available.

In chapter 2, I tell you about solar power—power from the sun—and how this is trapped and used. It can be a simple matter of building a house to the right design. Alternatively, special equipment can be built into the house to help store the sun's heat.

Chapter 3 is about the power of the wind. Wind is already being used to generate electricity in countries like the United States, Canada, and Britain. Look again at picture [1]. On the left is a wind-turbine—symbol of a new age. Do you know what a wind farm is? I tell you about wind farms and the different types of wind machines that are being built.

A turbine is a motor consisting of a vaned wheel made to revolve very fast by the force of a jet of water (or steam, or expanding hot air), or by the strength of the wind.

There are quite a few ways of using the force of water to generate electricity. Hydro-electric power stations and tidal power stations use the movement of water in rivers and estuaries. Scientists have ideas for machines which rock on the waves. Even heat contained in the waters of the sea need not be wasted. In chapter 4 I describe the work that is being done to harness all this energy.

We are still learning how to use renewable energy in the best possible ways. Can it slow down the rate at which we are using fuels? Can it ever replace them entirely? This is an exciting story of discovery, which all young people will share in before many more years have passed.

I · *What is energy?*

When you turn on a light in your house, do you ever think about the electricity that makes this possible? How is it made, and how does it reach your house?

It was an English scientist called Michael Faraday, in the nineteenth century, who discovered that moving a wire through a magnetic field produced an electric current. Faraday did this by turning a loop of wire around a magnet. The same result is achieved by keeping the wire still and moving the magnet.

An electric current is a controlled flow of electricity.

Faraday was using the energy from his own body to turn the wire around the magnet, and only a small amount of electricity could be produced.

Power stations, like the one in picture [2], work on the same principle of moving a magnet around a wire, but they use a machine called a generator to produce huge amounts of

[2]

[3]

electricity. The magnet in the generator has to turn at very high speeds, so a great deal of energy is needed to turn it.

In simple terms, energy is heat. Almost all the energy on the earth comes from the sun. The heat from the sun keeps our planet alive. It makes plants grow. It heats the air around the earth, causing winds. It warms the waters of the sea. Waves and currents are created by the wind and the sun. Without the sun, Earth would be a block of ice with no life on it.

There are two sorts of energy—stored (or *potential*) energy, and active (or *kinetic*) energy.

Potential energy has to be released from the object it is in. The food we eat contains potential energy. Our bodies burn the food and turn it into the energy we need to walk, run, and even breathe. Fuels make machines work in the same way that food makes us work. When the fuels are burned, the stored energy is released.

Kinetic energy is energy contained in a moving body. When you throw a ball through the air, the energy you put into the throw is passed to the ball. When the energy you have given the ball runs out, it falls to the ground. Anything which moves has kinetic energy. The wind moves the air, and the waves move the waters of the sea. Wind and waves have kinetic energy. This energy is not released by burning. It has to be *harnessed* in special ways.

Picture [3] shows the coal stocks at a power station. Other power stations use oil. Most of the energy for making electricity comes from burning vast amounts of coal and oil. The heat from the burning fuels boils water to make steam. This steam rushes through pipes to strike the blades of a turbine, making it spin at very high speed. The turbine is connected to the generator by a shaft. On the generator shaft is a magnet surrounded by copper bars. When the turbine spins, the magnet also spins, and the spinning magnet generates electric current in the copper bars. This electric current is carried away along transmission lines like those in picture [4].

[4]

What's wrong with using fuels?

If, by burning fuels, we obtain all the electricity we need, why think about other methods?

There are two reasons. One is that the fuels we are using will run out one day. The other is

[5]

Pollution (in the sense it is used here) means dirtying and fouling the landscape in a way which might cause lasting damage to it.

danger from pollution.

Oil, coal, and gas, are called *fossil* fuels. They were formed millions of years ago, long before people lived on Earth, by the decay of plants and animals in seas, rivers, and swamps. As these plants and animals died, they sank into sand and mud and were gradually buried deeper and deeper beneath further layers. Heat and bacteria worked on them and turned them into oil, or coal, or gas. Today, we are taking these fuels out of the ground as fast as we can. This is what the production platform in picture [5] is doing. It is pumping out oil from beneath the sea bed, to send it to be burned in a power station.

There are huge deposits of coal, oil, and gas in the world, but it took millions of years to produce them. By the time we have used up all the fuels available, there will be nothing to replace them with for another few hundred million years.

People are also worried about the pollution caused by burning fuels in huge amounts. They give off smoke and dirt and gases and soot which damage buildings, cause fogs, and injure people's health.

A cleaner type of power which can be used is nuclear power. There are now nuclear power stations which produce electricity by splitting atoms to release energy. You can see from picture [6] how clean a nuclear power station is, but some people feel that these power stations are not safe. There could be leakages of radio-active gas. Waste from nuclear power stations is radio-active, and getting rid of this waste is difficult. There is always a chance that some will escape into the atmosphere, so safety precautions have to be very strict.

Although some people are not happy about nuclear power, at the moment it is the only source of energy we have apart from burning fossil fuels. This is why more and more work is being done to find ways of using renewable energy sources. They are called 'renewable' because they will not run out. There will always be sunshine, wind, and water. Of

[6]

course, they are also unpredictable. No one can be sure that the sun will shine each day, or that the wind will blow strongly enough to turn a turbine. At the moment, the amount of electricity generated from these sources is small. But they are safe and clean, and they are free.

Alternative technology centres

Several countries have places where renewable energy sources are studied. For example, there is the National Centre for Alternative Technology in Wales, which provided some of the pictures for this book. You can visit these centres and see for yourself the sort of machinery that is being built to harness the power of the sun, wind, and water. Different sorts of wind machines have been built so that people can study which ones are the most efficient, and there are buildings which are heated by the sun, using the sort of equipment I tell you about in the next chapter. Electricity is also generated by water turbines. The work carried out at centres of this sort is fascinating. If the energy from these renewable sources can be harnessed and stored efficiently, they could be an answer for the future.

2 · *Power from the sun*

Solar power—which simply means 'power from the sun'—works in three ways. The most straightforward method is to make direct use of the heat the sun provides. This is called *passive* solar heating because it does not involve anything more than trapping and storing the sun's energy. *Active* solar heating requires specially designed equipment to trap and circulate the heat more efficiently. You can see a display of various solar heating equipment in picture [7]. Sunshine can also be converted into electricity by using solar cells called *photo-voltaic* cells. All three ways are explained later in the chapter.

The problem with any form of solar energy is that, although the sun is always there, it is not always shining. In some countries there are

[7]

long periods without much sunshine. In the winter the sun may not shine much, and when it does there is little heat in it. There are days when Earth's atmosphere blocks out the sun behind cloud and rain. Only in countries round the equator can you be sure of regular sunshine.

However, even in countries like Britain and Canada which have cold winters, work has been done to find ways of trapping sunshine. This is not a new idea. The ancient Greeks used passive solar heating systems to heat their houses 2,500 years ago. They built their houses facing south to catch the hottest rays of the sun. Large windows help to let in as much sunshine as possible. This is the simplest way of using the sun's heat.

Passive solar heating

In a passive solar heating system, sunlight is trapped and stored, and then circulates around the building.

If you go into a greenhouse you can feel that the air is hotter than the air outside. A car left parked in the sun can become uncomfortably hot. This is because the sun's rays shine through the glass windows and become trapped inside. They are absorbed by the materials inside—the seats of the car, for example—and converted into heat. That is why greenhouses and conservatories are built of glass to let in as much sun as possible.

A house heated by passive solar power works on the same principle. No special equipment is needed, just a sensible design. In the northern hemisphere, the sun is at its hottest in the

[8]

south. To make the most of this hot sun, houses are built with large south-facing windows. The buildings in picture [8] have been cleverly designed to trap the maximum sunlight. Inside the house, the walls and floors are designed to absorb heat. Dark colours absorb more heat than light colours. In hot countries, houses are painted white to keep them cool. To trap heat, walls and floors must be made of dark materials.

During the day, the sun shines through the large windows and its heat is absorbed by the dark materials in the house. It is stored there until nightfall, when the rest of the air in the room cools down. The warm air is then released from its store into the cool room. Hot air always rises, so the warmest air travels up to the ceiling. The cool air at the bottom is heated by the warm floor and rises to let in more cool air. So the air in the room is circulated by natural methods.

In most countries, there is not enough sunlight to make this an efficient way of heating a house on its own. Therefore houses have been built using active solar heating methods.

Active solar heating

Active solar heating systems use special solar collectors to trap and store heat. The heat is then distributed around the house by pumps and fans. There are two sorts of solar collector. One is called the flat plate collector, better known as the solar panel. This is a radiator-like panel with a glass top. The panel is made of a material which absorbs heat well, such as copper painted black. The sun's rays pass through the glass and are absorbed by the black panel. The panel is insulated to prevent heat escaping. Water can be heated by circulating it through pipes in the panel. This hot water is then pumped into a storage tank, or round a central heating system.

[9]

On a summer's day, the sun is at its hottest when it is directly overhead. Solar panels fitted into the roof, as in picture [9], can collect a lot of sunlight at this time. Below the panel the air in the loft space is heated, and can be circulated round the house.

Another device for heating a house by solar power is called a focussing collector. A curved mirror concentrates the sun's rays on to a small point, producing a very high temperature. Have you ever burned holes in paper with a magnifying glass? On a large scale, this method

[10]

can be used to generate electricity. High up in the Pyrenees stands the solar power station at Odeillo in France. More than forty large reflectors direct the sun's rays on to a huge, saucer-shaped mirror. In turn the enormous heat which builds up on the mirror is deflected down towards a furnace of water, turning it to steam. The steam is then used to generate electricity in the way I have already described. Picture [10] shows the reflecting panels, mirroring the equipment, trees, and hills before them. In the cover picture you can see another view of the same scene.

The problem with solar heating is storing the

[11]

energy for times when the sun is not shining. A heat store is needed. The most usual method is to have a large underground tank to store hot water. Heat can also be stored in the rock under a house. Both these methods are expensive ways of storing energy.

Another way of generating electricity from sunlight is to use special cells made of silicon. These are called solar cells or photo-voltaic cells, and you can get an idea of what they look like in picture [11]. They turn the energy from the sun into electricity. All they need is sunlight falling on them.

Solar cells were developed in the 1940s, and today can power satellites and space craft. Until recently they were expensive to produce, and so were used only for specialized work where electricity could not be generated by other means.

3 · *Power from the wind*

No one is quite sure when the power of the wind was first used. Certainly, the Egyptians were sailing barges on the Nile 5,000 years ago. Simple wind machines were probably used for grinding grain in Persia 2,000 years ago. Windmills with cloth sails, similar to the one in picture [12], have been used in Crete for centuries to pump water. In Holland windmills with rigid wooden sails are still used in great numbers. In picture [13] is a line of them, pumping water from low-lying country into the canal which carries it away.

The sails of a windmill are connected to a shaft, which can move a pump up and down or turn a grinding stone. The wind can be used to turn a turbine and drive a generator in the same way.

Windmills have been used for some time to generate electricity. In the United States windmills were used to pump water on farms, and earlier this century farmers began to use them to generate electricity. However, these windmills produced only enough electricity for a couple of light bulbs and a radio. To

[12]

[13] produce electricity on a large scale, many huge machines are needed.

A kilowatt is 1,000 watts, and a megawatt is 1,000,000 watts.

Electricity is measured in watts, kilowatts, and megawatts. Small machines, like those on the farms, might produce 200 watts. A larger machine producing 10,000 watts (10 kilowatts) could supply enough electricity for two houses or one farm. But the machines which interest people at present are those which produce more than 1,000 kilowatts (1 megawatt). This would be enough power for a village of two or three hundred homes.

Several different wind turbines have already been invented. There are two main groups of them. One is called the horizontal axis machine, and the other is called the vertical axis machine.

The first type is the traditional one with sails or blades. The wind blows against the blades, pushing them round on a horizontal axis. The old-style machines might have as many as twenty blades. Wind-driven water pumps, like the one in the centre of picture [14], are still common on remote farms in Australia, Africa, and the United States. Machines with many blades or sails will turn in low winds, but they will not turn very fast.

Modern wind turbines for producing electricity are also called aerogenerators. Aerogenerators have fewer blades than a windmill—normally two or three—and a tail vane to keep them turned into the wind. Aerogenerators need a strong wind to start them off, but when they get going they turn much quicker than a machine with more blades.

In Britain there are several small wind-turbines already at work, and a giant one is soon to be built. Picture [15] is a photograph of the existing fuel-powered station in Kent, with an artist's impression of the turbine standing in front. This huge machine would stand eighty metres (260 feet) high and its double blade would have a span of ninety metres. A machine like this would produce 4,000 kilowatts (4 megawatts) of electricity. This would be enough to supply a community of 4,000 people.

There are several experimental wind turbines already operating in the United States. One is the huge turbine in North Carolina belonging to NASA (National Aeronautics and Space Administration). It began generating electricity in 1979, producing 2 megawatts.

[14]

[15]

The wind must be blowing strongly to generate electricity. The turbine has to turn at high speed, so a gentle breeze is not good enough. The wind speed must be at least 16 kph (10 mph). Picture [16] shows a wind-turbine on the coast at Carmarthen Bay, in Wales. When the blades are turning, they move so fast that the eye cannot follow them.

Vertical axis wind turbines look very different. You will see a typical one on page 2 of this book. They consist of two or three curved strips of metal attached at both ends to a vertical pole. They whirl round on a vertical axis, rather like an egg whisk. One of the advantages of this type is that it does not need to be turned into the wind to work efficiently. It is cheaper to build than a horizontal axis machine, and can generate electricity in lighter winds.

One of the world's largest vertical axis wind turbines is in Canada. It whirls around on Magdalen Island in the Gulf of St Lawrence, producing 250 kilowatts of electricity, which supplies fifty houses. Canada has plans for an even bigger vertical axis turbine which will stand 110 metres high and produce nearly 4 megawatts of electricity.

Wind farms

But even these large wind turbines can still supply only enough power for a small community. How could wind power ever take over from the fossil fuels?

The only way is to have groups of these turbines, sited in places with good wind speeds such as hilltops and coastal areas. In other words, 'wind farms'. There are already wind farms in California, in the U.S.A., and one of them is shown in picture [17]. California is a good place for wind farms. It has mountainous regions and high-speed winds. It also has the space to install the machines. It is estimated that wind farms supply about 150 megawatts of California's electricity. The United States has plans to produce more and more electricity using wind power.

But could wind farms ever replace power stations? One problem is that, even if you used the biggest wind turbines, about 1,000 would be needed to provide the same amount of electricity as a large power station. The turbines cannot be too close together. Horizontal-axis machines with a 90-metre blade span would have to be spaced one kilometre apart, so large wind farms take up a

[16]

[17]

lot of room. It would be difficult to find the space for them in many European countries, or near capital cities anywhere. The land in between the machines could be used, but who would want to live near a wind farm!

One suggestion, for countries with plenty of coastline, is wind farms out at sea. Large wind turbines could be built in clusters of 400 in suitably shallow seas. Imagine twenty-five clusters of 400 turbines each, with each cluster producing 1,000 megawatts. A wind farm of this size would provide 40 per cent of Britain's electricity, for example, and would save 40 million tonnes of coal or 24 million tonnes of oil. Many heavily industrialized countries of Europe would enjoy a similar saving. Australia, New Zealand, and the coastal states of the U.S.A. could be supplied with nearly all their electricity needs.

4 · *Power from the sea*

Much of the earth's water is always on the move. Rivers flow down to the sea. Waves caused by the sun and the wind break on to a beach. Tides in the sea, created by the moon and the sun's gravity, pull the sea away from the shore and then send it back in, twice a day.

Another sort of movement in the sea is a current. Currents are fast-flowing streams of water caused by the wind and the movement of cold and warm water. Strong currents are dangerous to swimmers, for they are powerful enough to pull a person out to sea.

Like wind power, the energy from moving water has been used for centuries. Water mills for grinding flour, like the one in picture [18], relied on the flow of a river to push the water wheel round and turn the millstones. Today these mills are not much used, but that same water power could be used to generate electricity.

The simplest type of water wheel is the undershot wheel. The wheel is built with its lower part resting in the river, and is turned backwards by the flow of the water. The overshot wheel is more efficient. This type of wheel has buckets all around its edge. It does not rest in the river. Water from the river is directed along a chute which ends above the wheel. As the water falls from the chute it drops into one of the buckets, as you see in picture [19]. As the bucket fills up, its

[18]

[19]

[20] increased weight causes it to move downwards. Immediately another bucket takes its place under the falling water; fills up, drops down, and so on. The weight of the full buckets turns the wheel.

This is the principle used today to generate electricity in hydro-electric power stations. Rushing water falls on to the blades of a turbine, and makes it turn. However, a very powerful thrust of water is needed to turn a turbine fast enough. The natural flow of a river is too slow. At a hydro-electric power station like the one in picture [20] which is in the north of Scotland, a dam holds back the river. The volume of water builds up and forms a reservoir. When the gates in the dam are opened, the water surges through. The force of it can turn the turbine fast enough to generate electricity.

Waves power

Waves have an up-and-down movement that could be used to turn a turbine if all the energy could be harnessed properly. At the moment, devices for using wave power are only at the planning or testing stage. There are two sorts. One type floats on the surface, and the other type is anchored to the sea bed.

One example of a floating device can be seen in picture [21]. It is called the Salter Ducks, and was invented by the British scientist Stephen Salter. It is a series of vanes, shaped (if you have a bit of imagination) rather like ducks. These ducks are connected to a hollow spine. The rocking movement of the waves makes the ducks bob up and down to operate pumps inside the spine. These pumps would push a high-pressure jet of water along a pipe to turn a turbine. The electricity generated would then be taken to the shore by undersea cable.

[21]

[22]

A hydraulic machine is one which is operated by a flow of water, or makes water flow. Hydraulic pumps are operated by water.

Cockerell's Rafts is another floating device, of which picture [22] shows models one-tenth the size. It was designed by another British inventor, Christopher Cockerell. A line of rafts is placed at right-angles to the waves. Between the rafts are hydraulic pumps. The motion of the rafts on the waves operates the pumps. Fluid or air is pumped at high pressure to drive a generator.

The Oscillating Water Column in picture [23] also floats on the surface. Water moves up and down in a box. This movement forces the air in the box out through small holes. The high-pressure jets of air can operate a turbine.

Devices which are anchored at the bottom of the sea also use the up-and-down motion of the waves. One example is the Vickers Duct.

[23]

Water goes up and down in a tube. The pressure in the tube squirts out water which can drive a turbine.

So far, only small scale models of these wave devices have been built and tested. Huge and expensive machines would be needed to supply a useful amount of electricity, and so the electricity would cost more than people can afford. Various alternative designs are being tested in Britain, Japan, and other countries to see if electricity can be produced at a reasonable cost.

Tidal power

Tidal power has been used to operate water mills since the Middle Ages. To make use of the tide, the mill needs a head pond at the high tide level. The water flows in at high tide, and is trapped in the pond as the tide goes out. When the tide has gone out some way, the water in the pond is released. It rushes back to the lower level, turning the mill wheel as it goes.

High tide is the time at which the tide reaches farthest up the shore, and low tide is the time at which it has retreated its maximum distance from the shore.

A tidal power station follows the same idea, but on a much larger scale. Tidal power stations will work only where there is a big difference between the high and low tide levels. There are not very many places in the world which are suitable. The first tidal power station began generating electricity in 1968. It is situated at the mouth of the River Rance in France, and makes use of the Atlantic tidal wave which surges into the estuary. Across the mouth of the river stands the dam shown in picture [24]. At high tide the sluicegates on the dam are opened, and the water flows into a reservoir. When the reservoir is full the gates are closed, leaving the water trapped at the high tide level. The tide starts to go out, but the gates are not opened immediately. When the tide has been going out for about three hours, and there is enough difference in the two levels of water, the dam is opened. The water from the high level surges out through the dam, turning turbines and generating electricity.

A sluicegate is a door which controls the flow of water in a sluice (an artificial water channel), or in a dam or similar.

Canada has some of the highest tides in the world, in the Bay of Fundy in Nova Scotia. The difference between high and low tide levels is 17 metres. A massive tidal power station is planned for the Bay of Fundy. A pilot project has been set up on the Annapolis River which flows into the bay. This power station started

[24]

to generate electricity in 1984.

There is a tidal power station at Murmansk in Russia, and Britain is looking at plans for a scheme in the Bristol Channel.

Solar power in water

Electricity can be generated by making use of solar energy in the sea. About 70 per cent of the world is covered by sea, so much of the sun's heat which lands on Earth falls on water. The idea of using solar energy from the sea was first thought of about a hundred years ago by a French scientist called Jacques Arsène d'Arsonval. Today, scientists have called it Ocean Thermal Energy Conversion (OTEC).

Picture [25] gives an artist's impression of an OTEC plant built on a platform at sea. It has nothing to do with rushing water, but makes use of the different temperatures of the water-levels in tropical seas. If you go swimming in the sea, you have probably noticed that the water on the surface is warmer than the water underneath. The deeper you go, the colder it gets. This is because the heat of the sun collects mainly on the surface. In tropical seas, the surface water is hot enough to boil some liquids, while the deep water is very cold.

One liquid which boils at a low temperature is ammonia. Liquid ammonia is pumped through pipes on the surface. The pipes have been heated by the warm water, and the ammonia boils. Gas from the boiling ammonia turns a turbine. Then the used gas is sent into

pipes which have been cooled by cold water from the depths of the sea. The gas cools, turns back into liquid, and is then pumped back to the surface to be reheated. The ammonia circulates round in this way to keep the generators running. The electricity would be sent ashore by undersea cables.

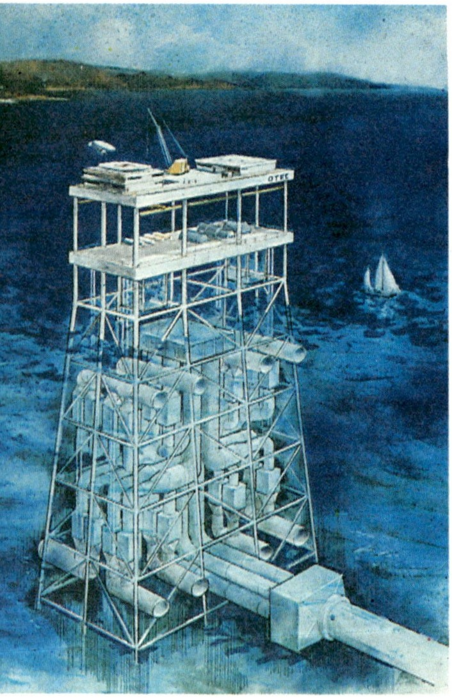

A mini-OTEC plant is being tested in Hawaii to see if the idea would work on a large scale. There are other places in United States waters where large OTEC platforms could be built. Countries like Britain and Canada cannot use this idea because their seas are not warm enough.

It is not only in ocean waters that the sun's heat is trapped. Heat is held in rocks below the surface of the earth. This heat from inside the earth is known as geothermal heat, and in some parts of the world rocks far below the surface have actually been melted by geothermal heat.

[26]

If water seeps down from the surface it is heated by these molten rocks. Pressure then forces the water back upwards through cracks and fissures, and it shoots out of the surface as hot water or as dry steam.

This water or steam can be used to drive turbines. The world's largest geothermal centre is the Geysers in California, which supplies electricity for homes in the area. Picture [26] shows a geothermal power station at work in Kenya.

I hope you have enjoyed reading about some of the ways in which we are already using renewable energy, and in which we plan to use it in the future. There are huge amounts of energy waiting to be harnessed. Unless we can use them effectively, we are not only wasting a valuable natural resource, but endangering our very future on this planet.

Index

Acknowledgements for photographs: Jacqueline Dineen, title page; Central Electricity Generating Board, 1, 2, 3, 4, 6, 15, 16; Documentation Francaise, picture on cover, 10, 24; Energy Technology Support Unit, Harwell, 21, 22, 23; National Centre for Alternate Technology, Wales, 8, 9, 11, 12, 14, 19; North of Scotland Hydro-electric Board, 20; Shell photograph, 5; U.S. Department of Energy, Washington, picture on Contents page, 7, 17, 25; World Wildlife Fund, 13 (Jill Matthews), 18, 26 (Mark Boulton).